Nofunland
HAD A PARTY

By John A. Honeycutt

Illustrations by Liljana Stojanovic

Another Hare-Brain Science Tale

Nofunland
HAD A PARTY

Another Hare-Brain
Science Tale
By John A. Honeycutt

Illustrations by
Liljana Stojanovic
Art Direction by
Kristina Ilievska
Production by
Layne Petersen

With love to:
Brady, Brad, & Matt

Way back,
a long time ago,
was the land
of Nofunland.

There was no such thing
as having fun in this land.
It was kind of a weird time,
and kind of a weird place.

Nofunland had six villages. When kids got old enough to choose, they could choose their own village.
They would live and work in that village the rest of their life.

6

That's just the way things were.
No one ever questioned why.
Each village had something
unique about it.

Village 1 was built
on flat ground, with two levels.
The lower level is where
the food and crops would grow.
The higher level is where
the houses were built.

Each morning, the people would wake up and go to work. They stood on a special device to be lowered from the higher level to the lower level. At the end of the day, the device lifted them from the lower level back to the high level.

Today, most kids
would recognize the device
as a teeter-totter.

But back then, it was just a
way to get from home to work
and back home again.
The villagers called it
a lever.

Village 2 was kind of like the first village, except the two levels were much farther apart.

The villagers connected the higher level and lower level with a long, narrow piece of aluminum.

13

Today, most kids would recognize
the device as a slide.

But back then, it was just a way
to get from home to work
and back home again.

The people would sit
at the top, and slide down
to work.

Then they would
climb up to go home.

The villagers called it
an inclined plane.

Village 3 was much different than the first two. All of the homes and businesses and stores were at the same level. No device was needed to get from home to work. 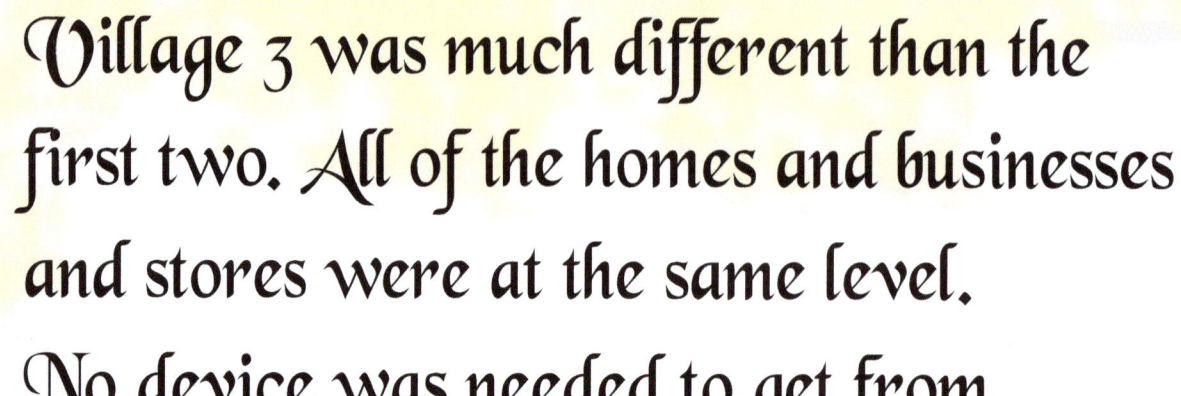 But this village did have a device that was used by all the workers.

It was the only type of device anyone used. The device came in several sizes. Sometimes it was used to chop down trees.

Smaller versions were used to slice through cheese. In fact, the village was well known for its timber and sliced cheeses.

Today, most kids would recognize larger version of the device as an axe. The smaller version was really just a kitchen knife. But back then, no matter how big or small,

it was just called a wedge.

Village 4 was just down the road. Village 4 used a device we now call the screw. When Village 3 cut down the wrong tree, Village 4 would try to repair it. They would try to screw the tree back together. Sometimes it worked, and sometimes it didn't. No one ever tried to repair any of the cheese.

Village 5 was very long and narrow. The homes were several furlongs away from the farmland and work places.

Village 5 used a special device to help them carry loads from one end to the other.

Today, most kids would recognize the device as a wheel.

By using a wheel on a cart, the workers could transport crops much easier.

Village 6 had two levels, sort of like 1 and 2. The levels were so far apart that a teeter-totter or slide would not have been practical.

Today, most kids would recognize the device they used as a pulley.

Even today,
rock climbers still use
a pulley
to climb up
and down a cliff.

The people in
Village 6 used
the pulley. The pulley
helped them climb
up to their homes.

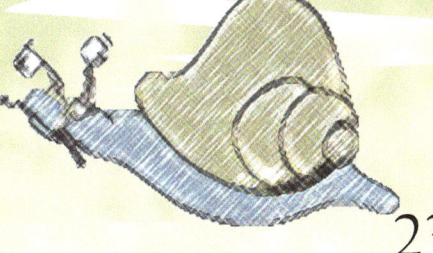

The Villagers did their jobs. Each Village used their device to help with their work. The lever, inclined plane, and pulley helped lift and lower people to their work.

Pulley

Inclined plane

Lever

The wedge
could cut down
timber and slice up cheese.

The screw could put
things back together.

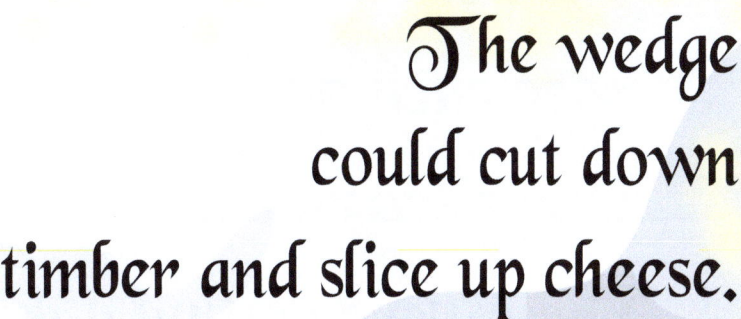

And the wheel could
help transport things
for miles.

Wedge

Screw

Wheel

Then one day, six kids happened
to see each other.
There was one kid from each Village.
They were getting a drink of water
from a stream that connected
all the Villages.

They said "hello" to each other.
One of the kids accidently
fell into the stream.
It was funny, and all of the kids
laughed - especially the one
who got
soaking wet.

Soon, the other five jumped in.
It was kind of cold.
Then the first kid said:
That was Funland!
The other kids agreed.

They jumped in
and out and had lots of fun.

Afterwards, they each went back
to their Village.
They did not see each other again
for several years.

The kids were grown up by now.
They were old enough
to choose their own village.
All the people from the six villages
gathered at the stream to see who
would select
their village.

The six kids were grown up by now.
They recognized each other.
Suddenly, they all rushed together
and jumped into the stream.
They laughed, and splashed,
and got soaking wet.

This baffled the other adults.
Then, one-by-one,
everyone got into the water.
Some only dipped in their toes.
Others dunked themselves
completely
under.

Soon, everyone was laughing.
Everyone was having
a good time.
They started chanting
Funland,
Funland,
this is Funland!

33

The Villagers realized that they should get together more often. They started working together, and communicating. Before too long, they started sharing ideas about how to improve their devices.

And they started coming up
with new ways
to make use of their devices.
All of the villages
started finding
creative uses
for all
six devices.

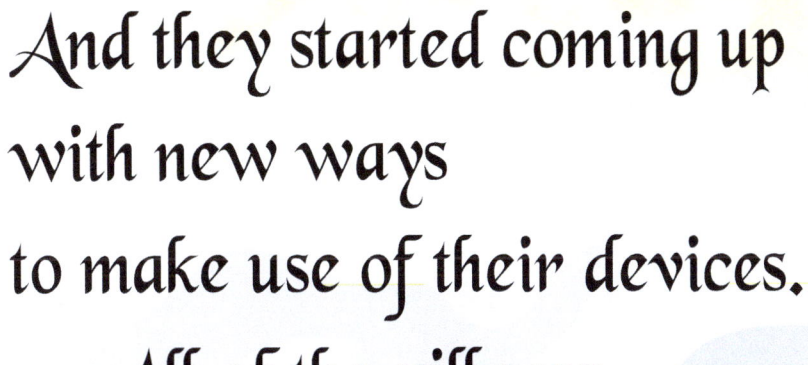

Some of the uses were practical and helpful. Other uses were fun and entertaining.

Over time,
complex combinations
of the devices were made.
People came up with
new machines,
new equipment,
and new ideas.
This made for
an exciting time.

Each year, Nofunland
had a party at the stream.
Eventually, the villagers
started calling the land
Funland.

One year,
at a big celebration,
a sculpture was
unveiled.

The artwork was kind of crazy-looking.

It had a lever, an inclined plane, a wedge, a screw, a wheel, and, of course, a pulley.

40

So, that is the story
of when
Nofunland
had
a party.

Lever

Inclined Plane

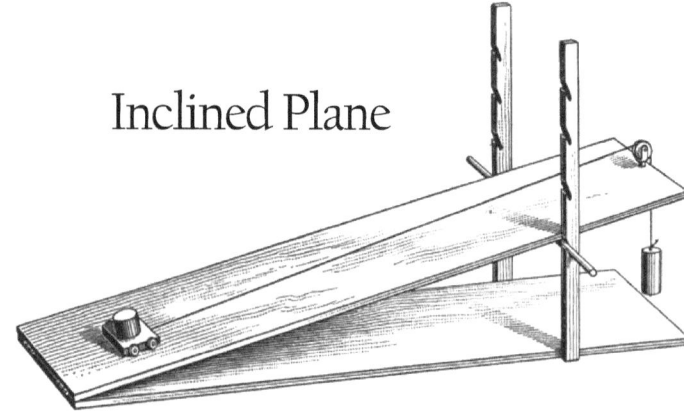

Wedge

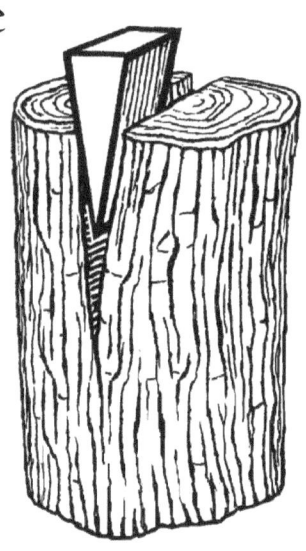

Screw

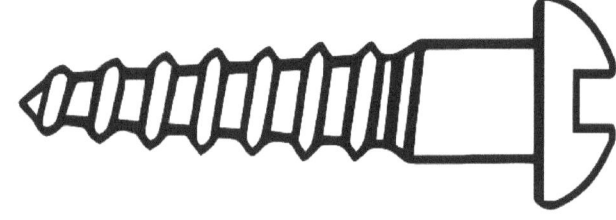

Wheel

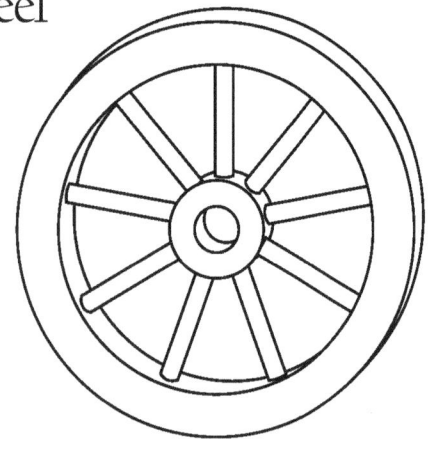

Pulley

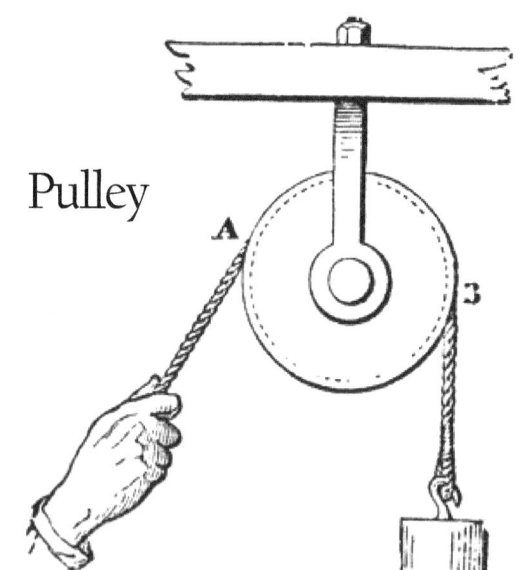

42

Simple Machines

Early Civilizations

Ancient civilizations made use of six simple tools to do their work. Young scientists can memorize this list:

- Lever
- Wedge
- Wheel and axle
- Inclined plane
- Screw
- Pulley

For example the ancient Egyptian societies used these simple tools to help construct the Great Pyramids (approximately 2700 BC to 1500 BC). Even before this time period, people used simple tools for hunting, farming, and gathering water. These six simple tools have been used by people and societies around the world for a long, long time.

NOTE: Scientists define a simple tool as a mechanical device that changes the direction or magnitude of a force.

Renaissance Era

Between the years 1300 AD and 1700 AD several cool inventions were made. Young scientists can have fun researching the cool inventions from this time period. Some of these inventions are surprising. Some examples include:

- The first mechanical clock
- The first steam pump
- The first submarine
- The first printing press
- The first flush toilet

All of these inventions make use of various combinations of the six simple tools.

Industrial Revolution

More recently, during a time period called The Industrial Revolution (1800 AD – 1900 AD), many more sophisticated machines were developed. Instead of people crafting everything from hand production, machines were invented to perform much of the work. This was the age of machine tools (tools that make other tools). It was also the time when machines were invented that made parts for other machines, including interchangeable parts. Something called "the assembly line" was invented during this time period. We still use assembly lines today. This speeds up the factory production of various products.

Young scientists can research interesting facts and inventions from this time period. Some examples include:
- The design of practical internal combustion engines
- The telephone, typewriter, and sewing machine
- The rapid growth of railways and steam ships

Modern-Day

Modern societies benefit from the inventions and ideas from earlier generations. Many times, inventors make improvements to earlier inventions. Here is a list of some of the cool things young scientists might hear about today in the news:
- Electric cars and hydrogen powered cars
- Significant increase in use of robotics
- Fully automated manufacturing
- Three-dimensional printing
- Hypersonic transportation

One day in the future, some of these newest ideas and newest inventions will seem old-fashioned. It is hard to imagine life today without the wheel, or the lever. But a long, long time ago, people were amazed by what a wheel and axle were able to do. Inventions like the printing press have changed the path of history. The combustion engine and telephone are so commonplace today, we almost take them for granted. The newest ideas like three-dimensional printing seemed impossible only a few decades ago.

NOTE: An interesting fact is that many of the newest, most cool inventions, still use combinations of the six simple tools described in this book.

Tomorrow

What new invention will happen tomorrow? Or next year? Or in ten years from now? Young scientists will be the inventors and engineers of tomorrow. What is it that YOU can imagine that hasn't yet been invented?

www.Hare-Brain.com

www.ingramcontent.com/pod-product-compliance
Lightning Source LLC
Chambersburg PA
CBHW040750200526
45159CB00025B/1839